STROKE AND APHASIA RECOVERY

Metaphors Help Us Mend

Thomas G. Broussard, Jr., Ph.D.

Johnny Appleseed of Aphasia Awareness

Stroke Educator, Inc.
Brunswick, Maine

The information and methods in this book are based on the author's knowledge, experience, and opinions for educational purposes.

Book and cover design by Thomas G. Broussard, Jr., and Sagaponack Books & Design

ISBN: 978-1-7344142-3-3 (softcover)
ISBN: 978-1-7344142-4-0 (hardcover)
ISBN: 978-1-7344142-5-7 (e-book)

Library of Congress Control Number: 2021925844

Summary: *Stroke and Aphasia Recovery: Metaphors Help Us Mend* includes images and aphasia-friendly definitions through a series of metaphorical stories about the brain and over 40 glossary entries describing stroke, aphasia, recovery, and neuroplasticity. The book educates people with aphasia, family and friends, caregivers, speech therapists, clinicians, doctors, and the wider public about this little-known language disorder.

MED057000-MEDICAL/Neuroscience

MED007000-MEDICAL/Audiology & Speech Pathology

SCI089000-SCIENCE/Life Sciences/Neuroscience

www.StrokeEducator.com

Stroke Educator, Inc.
Brunswick, Maine

To Laura,

Josiane, David and Will,

Mira, Kai, Fiona, Maddie, and Jack

A metaphor is a figure of speech
that directly refers to one thing
by mentioning another.
It provides the hidden similarities
between two different ideas.

Contents

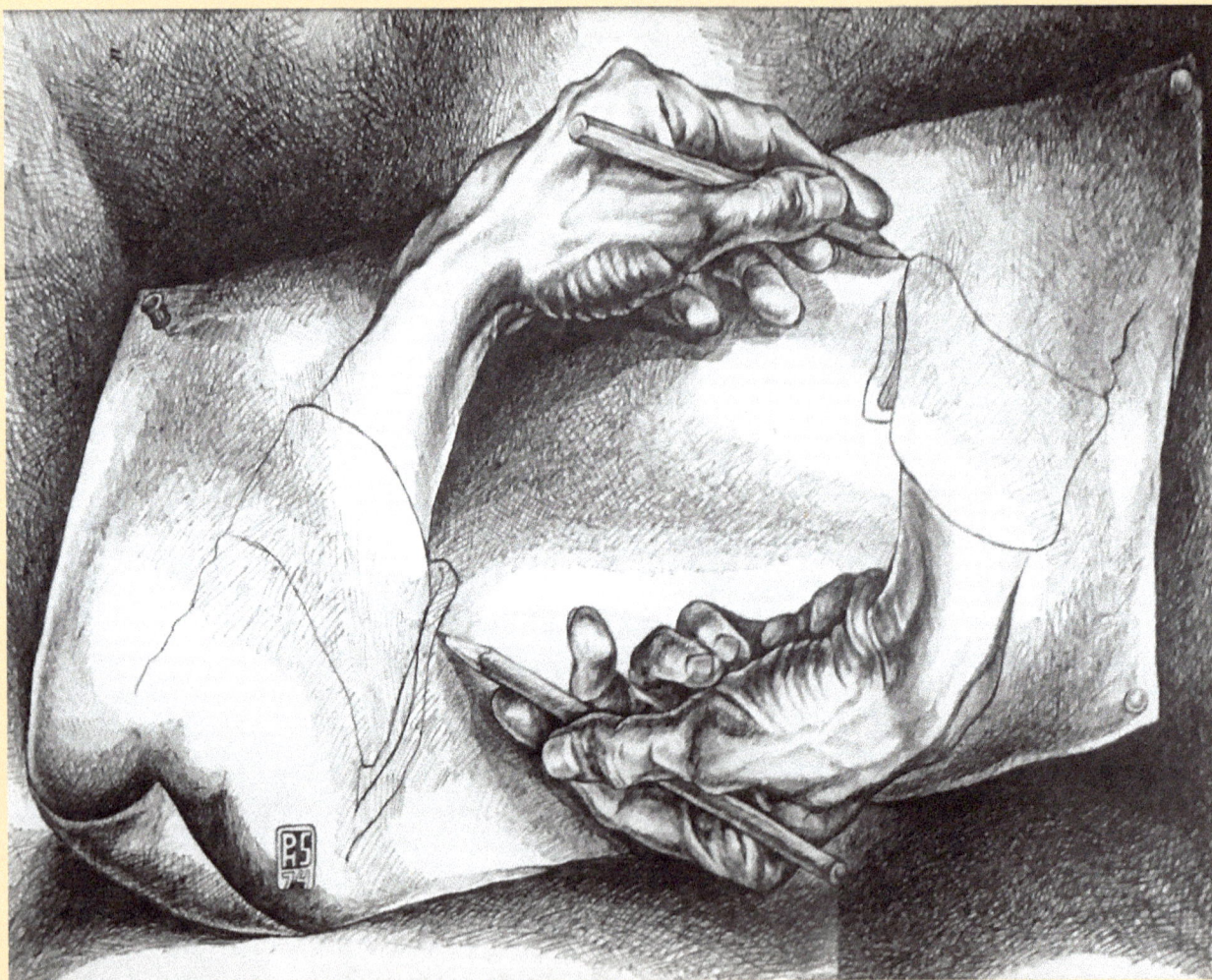

Hands Drawing Hands, by the late Patricia Houston Simpson, in the style of M.C. Escher.

Introduction

This is the second volume in a series of educational materials for people with stroke and aphasia, their family, friends, and caregivers, as well as the speech therapists and clinicians who care for them at the start of their lifelong journey towards recovery.

Aphasia is a communication disorder typically resulting from a stroke, with language problems including reading, writing, and speaking deficits. There are about 2.6 million people with aphasia in North America, and about 25% to 40% of people with a stroke acquire aphasia. It is more common than Parkinson's disease, cerebral palsy, or muscular dystrophy, but most people have never heard of it.

I had my first stroke and aphasia in 2011. I was an associate dean at The Heller School, Brandeis University, just outside of Boston, when I lost my language and could not read, write, or speak well. It took me years to regain my language. My mission now is to educate others about how the brain rewires itself—from the perspective of someone who experienced it firsthand.

Through a series of useful metaphorical stories in understandable language, this book helps us learn how the brain works and recovers. The mechanics of plasticity (the ability to change and create new brain matter) is the process by which we recreate our language. These lessons help provide you with the knowledge needed to generate the self-directed fuel that spurs plasticity, the foundation of all learning.

Stroke and Aphasia Recovery: Metaphors Help Us Mend is dedicated to helping people with aphasia, their family, caregivers, and clinicians, as well as educating the wider public about aphasia, as part of the international *"Aim High for Aphasia!"* Aphasia Awareness Campaign.

Tom Broussard
Johnny Appleseed of Aphasia Awareness

The Rosetta Stone of Aphasia Recovery: Persistent and Repetitive Language Activities

The "Rosetta stone" is a social meme that is considered the essential clue to understand the mechanics of any new field of knowledge. There is a Rosetta stone for aphasia recovery too: persistent and repetitive language activities, starting at the beginning of language therapy.

Persistent and repetitive language activities are the key to recovery, if provided daily for months, if not years, of practice.

Formal (and funded) therapy is just the tip (of the tip) of the iceberg of dedicated effort needed to rebuild the hundreds of miles of myelinated fiber and the connective tissue (dendrites and synapses) needed to regain one's language.

One issue is that conventional therapy provides brief, intermittent, and isolated bursts of therapeutic energy at the start, without the regular persistent and repetitive activities needed to maintain the momentum and to sustain plasticity for the long term.

Conventional therapy should include the basis for those activities long before formal therapy ends. Otherwise, you could be adrift without the tools needed to navigate to the distant shore of recovery.

Trim Tab:
Helping Steer the Ship of Aphasia Recovery

A trim tab is a rudder within a rudder. It takes a lot to turn a big ship with a huge rudder. It is so big that it often takes another rudder within the rudder. There is a tiny sliver at the edge of the rudder called a trim tab. It's a miniature rudder. Speech therapists are the important trim tabs of aphasia recovery.

Speech therapists must address both the short- and long-term therapeutic solutions for people with aphasia. The therapist's work is designed to start the process of repairing the language of people with aphasia.

Speech-language therapists must also establish an enriched environment by using speech and language activities to help create the habits needed for the long-term application of *those same activities*.

Repaired language does not occur overnight. It happens gradually, based on long-term, constant plasticity-induced pressure brought about by language activities within therapy, as well as outside of formal therapy.

Speech therapists help steer the ship and are the trim tabs of aphasia recovery. They provide the short-term startup. They must also work on supplying the tools needed for the longer journey home towards recovery.

Photosynthesis and Plasticity:
The Magic of Creating Something from Nothing

Photosynthesis is a process that converts sunlight into green leaves. Plasticity converts thought (it might as well be sunlight) and cognitive activities (reading, writing, and speaking) into neural (brain) matter.

Photosynthesis is a wonderful metaphor for plasticity, especially for people with aphasia (PWA), who regain their language through plasticity. Plasticity creates dendrites and synapses, the metaphorical branches and leaves of photosynthesis.

People are exposed to an enriched and therapeutic environment, but often have no idea that their activities provide the materials needed for recovery. One ingredient needed is regular and repetitive experience-dependent activity that provides the fuel to keep the machinery running.

But the machine doesn't develop on its own. It isn't enough to demonstrate one set of activities or another for a few minutes in any therapy session, and consider it to be therapeutic enough to induce plasticity for long-term learning.

The activities themselves must become habitual and deeply rooted, such that plasticity can continue long term, long after formal therapy has ended.

More sunlight, more photosynthesis; less sunlight, less photosynthesis. In much the same way, more thought and cognitive activities, more plasticity; fewer of those activities, less learning and plasticity. This is the direct link which literally rewires the brain.

Aphasia Découpage:
Applying One Coat of Learning at a Time

Découpage is the art of cutting and gluing paper, sealing it with varnish/paint, and repetitive sanding. Language improvement from aphasia uses repetitive language activities (reading, writing, and speaking) that paint (encode) and sand (consolidate) learning, one coat at a time.

Découpage is the art of cutting and gluing paper cutouts onto various items (small boxes, wooden handbags, furniture), in combination with paint and cutouts from magazines or cards.

As each of the cutouts is laid down, each layer is sealed with multiple coats of varnish. After the varnish coats dry, the surface is sanded before the next varnish is applied.

It often takes 30 to 40 layers of varnish and sanding for the "stuck on" appearance of the papers to disappear and begin to look like painting or inlay work with a polished finish.

The long-term process of improvement and recovery, when using repetitive language activities, demonstrates the progression of incremental learning at the cellular level—now sanded and stabilized for the next application of tomorrow's learning.

Repetition and Word-Finding Activities: Neural Knitting by Any Other Name

Ongoing word-finding activities set within an enriched environment provide the opportunity for plasticity to occur. Once activated, plasticity provides the habitual neural knitting that accomplishes more with less effort than before the previous attempt.

It is impossible to describe the *process* of how one's brain gets better *while* in the process of *getting* better. It is just as difficult to try to explain how knitting works.

Like many things, the process of learning is through experiencing the process itself. The process provides the learning of the product. Knitting is hugely repetitive and, once trained, knitters are unaware of the steps their hands are taking.

The brain performs many functions, like riding a bike or tying one's shoes, without any conscious choices regarding the steps. Once "locked in memory," the steps continue unabated.

Recovering one's language from stroke and aphasia is a repairable process using repetitive actions, just like knitting.

Language Repair:
Darning the Neural Fabric after a Stroke

Darning is a traditional sewing technique used for repairing holes in socks. It is often done with simple stitches that are woven into rows to fill up the hole or worn area. Similarly, language activities themselves are the neural darning, filling in the holes, gaps, and missing parts, and repairing the brain damaged by a stroke.

The definition of darning "restores the fabric to its original integrity" and is the common denominator between that and, metaphorically, repairing one's language after a stroke.

The brain grows new neural matter from the remaining cells, like sewing a patch in the brain. New threads are spun by plasticity (the capacity to grow neural matter) and are woven into the neural fabric, replacing the language function of the cells that were lost.

Repairing one's language is "darn" tough and requires *much* more work (and practice) with various language activities (reading, writing, speaking) that induce plasticity and sew up the holes and gaps in the language lost by a stroke.

Neural Polygraph:
A Duplicating Device

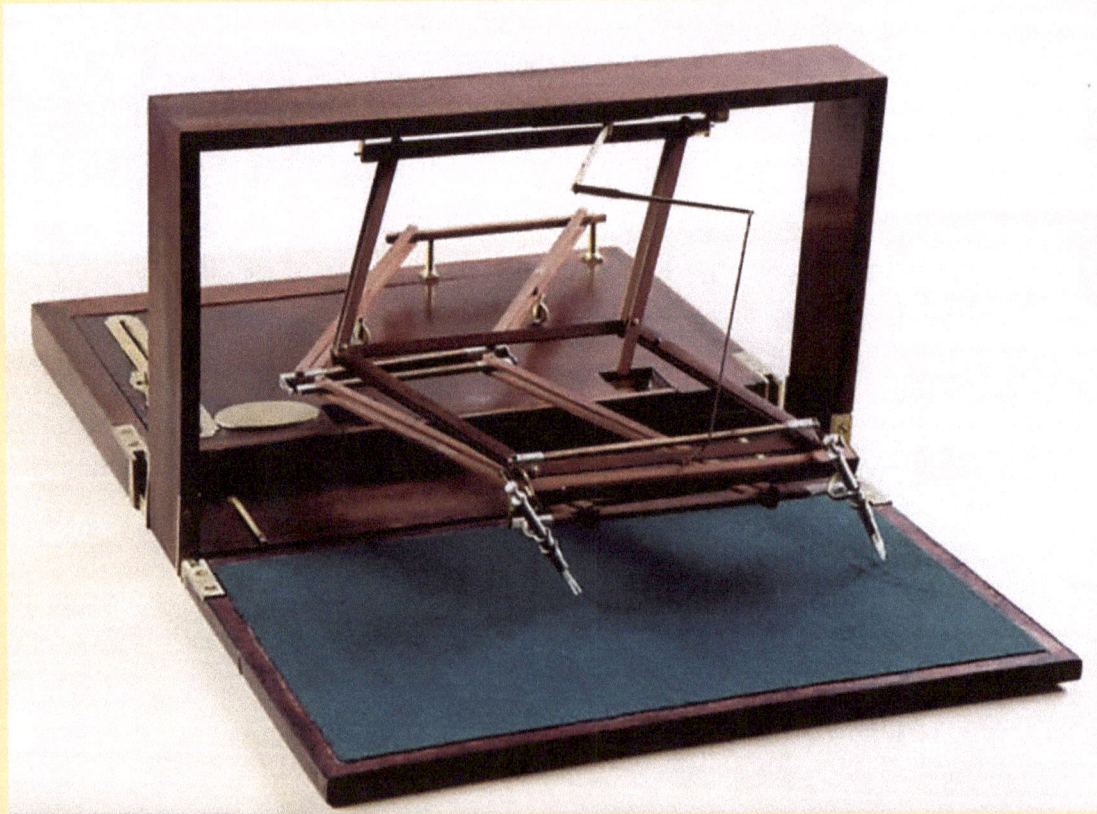

An eighteenth-century polygraph was a device that could copy a second set of letters. Intensive, consistent, and repetitive cognitive activities on the *outside* create neural copies on the *inside* with plasticity ink™.

President Thomas Jefferson used a letter-copying device called a polygraph (duplicating device) and copied tens of thousands of his letters. (A lie detector device is also called a polygraph.)

His hand moved one pen, which operated a second pen that produced a copy of the original letter. His polygraph is located at Monticello, in Charlottesville, Virginia.

Watching how the machine works, one begins to understand how cognitive and language activities, working on the outside, produce, copy, and create new neural building blocks of the same item or idea, on the inside. The polygraph is a metaphorical key for understanding how plasticity works.

People with aphasia can begin to understand the magic of plasticity and the allure of a *neural* polygraph with copying capacity.

Coppicing Trees:
Regrowing the Learning Field

Coppicing is a method of tree management based on the regeneration of new trees from a stump when the original tree has been cut. In the same way, the remaining cells, after a stroke, have the capacity to grow metaphorical branches and leaves, and regrow the learning field.

When a tree is cut down, a stump is left. Depending on the type of tree, young trees can sprout into larger trees from the original stump, resulting in many "generations" of trees from a parent tree.

In the same way, the remaining brain cells after a stroke (the metaphorical stumps) have the capacity to grow new branches (dendrites) and leaves (synapses) from the cells that are left.

Experience-dependent activities (reading, writing, and speaking) are the key to neuroplasticity's growth. Persistent and repetitive language activities are the active ingredients needed to grow and repair the brain.

It is in the nature of neurons (brain cells) to convert thought and cognitive activities into the fuel needed to feed the remaining cells. These cells are enhanced in their ability to regrow the thoughts of the previous thinking cells.

Playground Merry-Go-Round Spinner: Homework Is More Than Just *Homework*

Homework is more than just run-of-the mill *homework*. Homework for people with aphasia (as well as for others) requires a continuous push (like the playground merry-go-round) to keep and sustain the neural machinery spinning, from day one.

Daily language homework is necessary in order to acquire the habitual activities needed to develop successful lifelong learning and personal therapy.

We have the tools. We have to *use* them persistently, repetitively, and consistently to induce and maintain plasticity and the resultant growth and learning.

The elementary school playground's merry-go-round is a useful metaphor for aphasia recovery. Once the spinner is up to speed, it doesn't take much to keep it going.

A regular push keeps the speed up for much less energy and effort than was needed at the start. Homework for language activities provides the daily "push" for aphasia recovery.

Formal but short therapeutic activities induce plasticity that fades without the next "push." However, day-to-day homework maintains the ongoing stimuli (experience-dependent activities) and continuity that keep the merry-go-round of recovery going.

Russian Dolls:
The Nested Attributes of Aphasia and Recovery

The metaphor of a Russian nested doll (nesting within each other) helps us understand the concept of one difficult problem after another to be solved. The deficits of language for people with aphasia are similarly situated; various deficits are nested inside each other.

A stroke can strike anywhere in the brain and damage different modalities in different ways. The deficits which people with aphasia (PWA) deal with may seem imposing. And the deficits may go beyond being unable to read, write, or speak well. Additional deficits of one modality or another are often nested inside the main ones.

PWA are often unaware of their deficits, especially at the start. They become more aware of their deficits as they practice reading, writing, and speaking. Though awareness may be lagging in one language skill, the other skills can still improve.

Practicing allows more deficits to appear, even while the recently discovered deficits seem to be in the process of disappearing. The process of recovery includes the appearance and subsequent disappearance of various nested deficits.

"The Princess and the Pea": Small Things Can Make a Big Difference

Everyone knows the fairy tale "The Princess and the Pea", by Hans Christian Andersen, and how small things can make a big difference. In the world of aphasia recovery, this concept relates to the regaining of one's language after a stroke.

A prince wanted to marry a princess, but he needed a *real* princess. One night a young woman came to the castle looking for shelter, and claimed to be a real princess.

The prince's mother tested her by placing a little pea in the bed, covered by huge mattresses and twenty feather beds. In the morning, the young woman had not slept because there was something hard in the bed. She could feel the pea. It was a small thing, yet important.

Successful habits have been part of our lives for years, often without our knowing how or when they started. Habits can be the princess's pea for people with aphasia.

Small habitual activities and their resultant habits were created early in life. People with aphasia can continue to use those same activities *after* the stroke, as well, with no conscious idea that the activities were and continue to be highly therapeutic—even without formal therapy.

It turns out that the activities *themselves* are the active ingredients of language recovery and improvement for people with aphasia.

The same habits and activities that led to learning those skills the first time are the same habits to use to relearn reading, writing, and speaking.

Motivation:
The Little Engine That Could

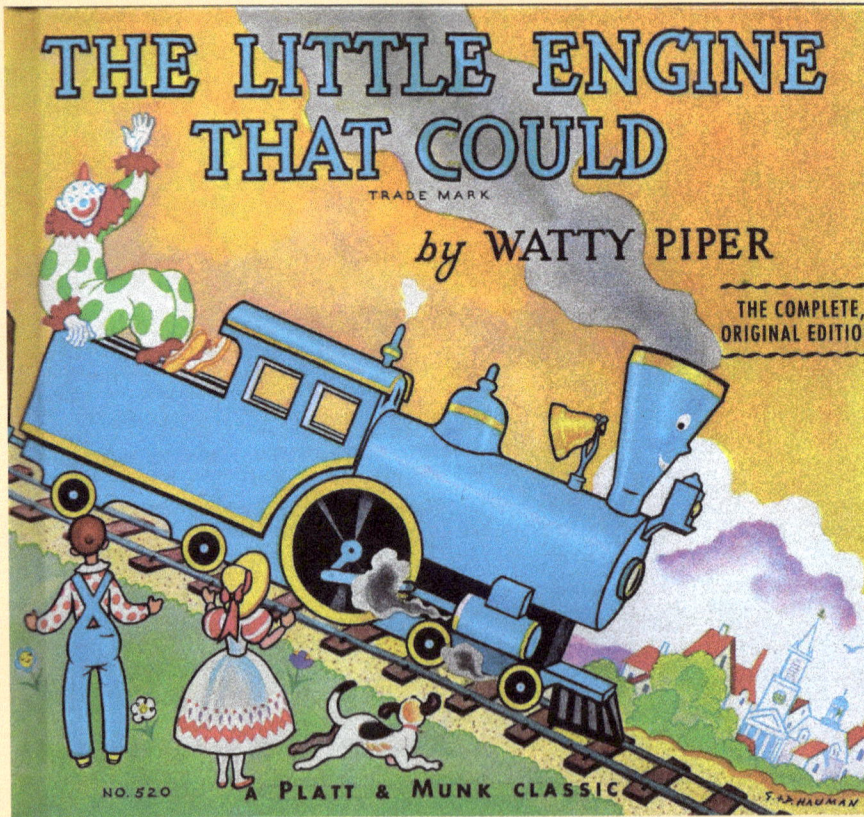

THE LITTLE ENGINE THAT COULD
TRADE MARK
by WATTY PIPER
THE COMPLETE, ORIGINAL EDITION
NO. 520 A PLATT & MUNK CLASSIC

The message of *The Little Engine That Could* is that it climbed the mountain that others couldn't, assuming the little engine believed that the "I think I can" thought was more than half the battle.

The Little Engine That Could is a famous story that teaches children the value of optimism, hard work, and motivation. In the tale, a long train is being pulled over a big mountain until its engine breaks down. Other engines are asked to pull the train, but for various reasons they refuse. Then a small engine agrees to try.

The engine succeeds, repeating its motto, "I think I can, I think I can," and then on the other side of the mountain, saying, "I thought I could, I thought I could."

The brain, not unlike the little engine, has incredible power even after it has been damaged from a stroke. The brain has the capacity to rewire itself and regain its language (depending on the severity of the stroke).

Recovering one's language from a stroke and aphasia can seem like an impossible task. However, remember the five rules of aphasia recovery: motivation, and practice, practice, practice, and more practice!

Experience-dependent neural plasticity is the engine, and persistent language activities provide the fuel: the more practice, the more healing and the more improvement. The activities *themselves* make the engine work.

The Illusion of "Reaching One's Plateau"

People with aphasia cannot *not* learn at any stage. That is the illusion of having "reached one's plateau" when there are still so many mountains to climb.

Recovery takes place using two different mechanisms that often occur at the same time: one by nature (and relatively fast) and the other by effort (and imperceptibly slow).

The first phase ends quickly once the inflammation, or swelling, has healed.

That leaves the second, and very slow phase, providing the *impression* that progress is slowing and coming to the end, when in fact it is a lifelong and ongoing feature of the neural mechanism of rebuilding the networks of the brain.

At this point, sadly, patients are often labeled as having "reached their plateau" by *comparing* the faster, spontaneous, and natural segment of healing to the subsequently slower process of creating new brain matter at the cellular level.

Once language improvement appears to have slowed, it makes it even more difficult to continue with formal (and funded) therapy under the current rules in place.

It is an impossibility *not* to learn, given that every single neuronal impulse is transmitting new information to other (and continuously growing) cells with the lifelong ability to learn. So, we continue to practice and learn.

Topiary:
Shaping the Networks of the Brain

Topiary is the art of sculpting, clipping, shaping, and trimming shrubs, plants, and trees into ornamental and fanciful shapes in indoor and outdoor landscaping. Similarly, experience-dependent neuroplasticity creates a "brain topiary" that shapes and trims the brain's neural networks.

Experience-dependent neuroplasticity uses cognitive activities (reading, writing, speaking) that provide the stimulation which enhances and sculpts the learning field of the brain.

The lifelong process of creating, organizing, and shaping the neural connections occurs as a result of a person's life experiences.

The brain is *built* to repair the living environment by using the same tools, the same activities, and the same habits (similar to shaped steel wire frames for topiary) that were used the first time the links were established in an enriched environment.

People can be *better* therapists and patients (or teachers and students) if they become more conscious and aware of how the brain's "topiary" works. As a result, people with aphasia can become *master* gardeners, weaving the mazes of their own recovery.

The Daily Alchemy of Neuroplasticity

In ancient times, alchemy was thought to be able to convert lead into gold—but such magic doesn't work in reality. However, neuroplasticity can appear to be magical as it converts thought and cognitive activities into brain matter and the resultant learning.

Clocks are designed to tell time and are physically linked to the second, minute, and hour hands moving at different increments of time. People can't see the movement of the minute or hour hand of a clock because it is below the level of visual perception.

Without one's understanding of the underlying cogs and wheels that make the neural clock work, much of the effort of speech therapy can give the appearance of being for little or no result.

Most people with aphasia learn that their activities are therapeutic, but there is no quick *cause* with an immediate *effect* as a result of just an hour of intermittent speech therapy.

Plasticity converts thought and cognitive activities into brain matter, but it takes time, repetition, and persistent effort to induce long-term, habitual, experience-dependent neuroplasticity and to see its satisfying effects.

Illustration Credits

"Hands Drawing Hands," by the late Patricia Houston Simpson, in the style of M.C. Escher. Patricia Simpson (1947–1999) was educated at the Corcoran School of Art in Washington, DC. In 1973, she married the author's Naval Academy classmate, Michael Simpson, and continued to work as a freelance artist throughout her life. She became well known for her lifelike animal portraits, as well as many humorous and fantasy paintings and drawings. – Collection of the author

Coppicing trees, Rosetta Stone, trim tab, découpage, darning egg – Collection of the author

Knitting, photosynthesis, topiary, merry-go-round spinner, wizard/alchemy, princess and the pea, plateau, Russian dolls background – Can Stock Photo

Polygraph – Wikimedia Commons

Russian dolls – Creative Commons

The Little Engine That Could, The Complete Original Edition, by Watty Piper (Platt & Muck Edition). Approved by Penguin Random House, 3/16/2021.

Thanks to the folks at Celebration Tree Farm & Wellness Center in Durham, Maine, for information on coppicing trees.

Aphasia Website Resources

- Adler Aphasia Center
  https://adleraphasiacenter.org/

- American Heart Association (AHA) / American Stroke Association (ASA)
  https://www.heart.org/

- American Speech-Language-Hearing Association (ASHA)
  https://www.asha.org/

- Aphasia Access
  https://www.aphasiaaccess.org/

- Aphasia Center of California
  http://www.aphasiacenter.net/

- Aphasia Center of Maine
  https://www.aphasiacenterofmaine.org/

- Aphasia Institute
  https://www.aphasia.ca/

- Aphasia Nation, Inc.
  https://www.aphasianation.org/

- Aphasia Recovery Connection (ARC)
  https://www.aphasiarecoveryconnection.org/

- Boston University Aphasia Resource Center
  http://www.bu.edu/aphasiacenter

- Brooks Rehabilitation Aphasia Center (BRAC)
  https://brooksrehab.org/locations/aphasia-center/

- Lingraphica
  https://www.aphasia.com/

Aphasia Website Resources (continued)

- National Aphasia Association (NAA)
  https://www.aphasia.org/
- Stroke Comeback Center
  https://strokecomebackcenter.org/
- Stroke Educator, Inc.
  www.strokeeducator.com
- Stroke Support India
  https://strokesupport.in
- Tactus Therapy
  https://tactustherapy.com/
- The League for People with Disabilities, Inc. / The Snyder Center for Aphasia Life Enhancement (SCALE)
  https://www.leagueforpeople.org/scale
- The Shirley Ryan AbilityLab
  https://www.sralab.org/
- Teaching of Talking
  https://teachingoftalking.com/
- Triangle Aphasia Project (TAP) Unlimited
  https://www.aphasiaproject.org
- UCF Aphasia House
  https://healthprofessions.ucf.edu/cdclinic/aphasia/
- Voices of Hope for Aphasia
  http://www.vohaphasia.org/

Glossary

aac (augmentative and alternative communication) devices. Devices that offer communication methods and machines used to supplement or replace typical communication modes (as in writing or speech) for people with aphasia (and other brain diseases).

active ingredients. Active ingredients used in drugs are biologically active. When it comes to aphasia, the active ingredients needed for recovery are found in the use of continual, repetitive, and persistent language activities (reading, writing, and speaking), plus exercise that induces neuroplasticity.

activities. Therapeutic "activities" for people with aphasia are *themselves* the active ingredients that induce and "turn on" plasticity and the resultant recovery. Speech therapy for each deficit *requires* action to regain lost skills. Therapy for writing *requires* constant writing. Therapy for reading *requires* constant reading, and so on.

axon. A nerve fiber; a long, thin "wire" of a nerve cell, or neuron. It conducts electrical impulses away from the body of the nerve cell. Its function is to transmit information to different neurons, muscles, and glands in the brain.

cell. See **neuron**.

cerebrovascular accident (CVA). Known as a stroke, it is an injury of the brain where the blood supply to a part of the brain is interrupted, either by a clot in the artery or an artery bursts.

conventional therapy. The beginning of language therapy which includes naming, repetition, sentence completion, following instructions, and conversations. Conventional language therapy (approximately 50 hours) is the start of helping patients relearn communication to lessen deficits (such as in writing, reading, and speaking) that were caused by stroke and aphasia.

dendrites. Small branched extensions ("short wires") of a nerve cell that deliver electrochemical stimulation and transmit information to other dendrites by upstream neurons. Dendrites play an important role in integrating the inputs before sending the newly integrated output to other downstream receptors.

enriched environment. The added stimulation of the brain by its physical and social surroundings. More stimulating environments lead to higher rates of (synaptic) growth (leaves), and more complex branching (dendrite arbors), which lead to increased brain activity, growth, and learning.

enriched therapy. Conventional therapy (50 hours) is the start to aphasia recovery; intensive therapy (150 hours) is a sprint; and enriched therapy (1,500 hours per year) is a marathon. These activities include persistent reading and writing, regular social interaction and speaking, plus consistent exercise and walking. The activities provide the lifelong communication (lifelong learning) marathon needed to induce plasticity and the resultant learning.

evidence. Keeping track and saving the results of various communication activities allows for reflection, assessment, and feedback of ongoing (and still changing) outcomes. Examples include: keeping and reviewing a diary, listening to your voice recordings, taking and viewing pictures, taking and watching one's video recording, and reviewing one's exercises.

experience-dependent neuroplasticity. Also known as brain plasticity or neural plasticity. The brain has the ability to change continuously throughout an individual's life. People with stroke and aphasia have the capacity to regain their language (and other functions), based on therapeutic (reading, writing, speaking) activities, plus exercise, all of which induce plasticity.

feedback. Occurs when outputs of a system are converted into inputs as part of a cause-and-effect loop of enhanced awareness and learning. People with aphasia need to establish a feedback loop for every language deficit in order to "see what you *wrote*," "hear what you *said*," and "watch what you *saw*." Without gathering and saving the evidence, there would be no instructive lifelong feedback loop, resulting in less learning.

habit. Usually an unconscious set of actions acquired through frequent repetition, such as, "It is a habit never to be late." People with aphasia still have many of the habits they acquired before their stroke. The existing habits, although still unconscious in nature, can be a huge influence on language activities leading to plasticity and recovery.

high-tech applications. See **aac**.

intensive therapy. Intensive aphasia programs usually provide about 30 hours a week, for four to six weeks, of individual and group therapy to people with mild to severe aphasia, to help with language deficits by using repetition and socialization.

language deficits. After stroke and aphasia, and after being told that you have "lost your language" may be the first realization that your language is damaged such that you cannot read, write, or speak well in various proportions. It is also possible that you may be unaware of your deficits.

lost cells. A stroke destroys a large number of brain cells (neurons), ranging from hundreds of millions to billions of cells, and hundreds of miles of myelinated fibers, depending on the severity of the stroke. Recovery comes from the remaining cells that have the capacity to grow the new connections that the lost cells took with them.

low-tech applications. Interventions or activities (that are still highly therapeutic), with little or no cost or difficulty, using diaries, journals, voice recordings, video recording, calendars, photography (taking pictures), card games, battery-powered games, computer games, flash cards, and more.

myelin fiber (or sheath). A fatty substance that surrounds nerve cell axons (the nervous system's wires) to insulate them and increase the speed of information of electrical impulses passing through the axon. The myelinated fiber can be likened to an electrical wire (the axon) with insulating material (myelin) around it.

neuron. A neuron, or nerve cell, is a cell that communicates with other cells via specialized connections called synapses. A typical neuron consists of a cell body, dendrites, and a single axon. A group of connected neurons is called a neural circuit. Neurons grow new dendrites and synapses to learn something new or reconnect the messages that have been destroyed by a stroke.

neurotransmitter. A chemical messenger that carries, boosts, and balances signals between neurons and other cells in the body. These messengers affect a wide variety of physical and psychological functions and are part of shaping thoughts and actions.

pictures. Taking pictures is another therapeutic component that stimulates thinking and learning. People with aphasia who are unable to fully tackle reading, writing, or speaking because of their deficits, can take pictures of the world around them and review the images repetitively, later at home. This provides more cognitive understanding with every single glance.

plasticity ink™. A metaphorical device of plasticity and the ability of the brain to change at any age, indicating that all learning, whether healthy or not, is written in plasticity ink™.

plateau. Described as a period of activities with little or no improvement or learning. People with aphasia are sometimes described by clinicians as "plateauing," when there is not enough progress or clinical improvement. However, the people *with* aphasia themselves never describe themselves as "plateauing," given that their lifelong day-to-day learning is the basis for improvement, no matter how small.

practice. The five rules of aphasia recovery are motivation, and practice, practice, practice, and more practice! The secret to recovery is *more* practice, by doing almost anything of interest that piques your curiosity and drives you to satisfy your need for constant study and reading. The resultant learning then urges even *more* practice and study.

principles of neuroplasticity.
The 10 principles include:
- Use it or lose it.
- Use it and improve it.
- Specificity.
- Repetition matters.
- Intensity matters.
- Time matters.
- Salience matters.
- Age matters.
- Transference (or generalization of skill or activity).
- Interference.

These principles contribute to plasticity and learning.

prior knowledge. People use their prior experiences and knowledge to learn something new. People with aphasia still retain their past relevant experiences with which to attack the new problems of the reading, writing, or speaking deficits. Knowing what you know and connecting new knowledge to old is the foundation for subsequent learning.

repetition. One of the principles of neuroplasticity. But repetition doesn't mean that you have to repeat a word over and over again. Saying a word, writing the word, then seeing a picture of the same item induces plasticity, which builds a neural copy in the brain such that one can remember the word better and faster with more integrated action.

salience. One of the principles of neuroplasticity. The more noticeable, important, or memorable an item is to a person, the more likely they are to remember the item or activity since it "stands out" among the many.

severity. The severity of a stroke is measured by the NIH (National Institutes of Health) Stroke Scale, which assesses several aspects of brain function, including consciousness, vision, sensation, movement, speech, and language. Stroke severity uses a scale scoring system: minor stroke (1–4), moderate stroke (5–15), moderate/severe stroke (15–20), and severe stroke (21–42).

stroke. See **CVA**. A medical condition in which obstructed blood flow to the brain results in cell death. There are two main types of strokes: ischemic (85%), due to lack of blood flow; and hemorrhagic (15%), due to bleeding. Symptoms of a stroke include an inability to move or feel one side of the body, problems understanding or speaking, dizziness, or loss of vision on one side.

stroke survivors. All strokes are different, as are all stroke survivors. The impairments of stroke survivors (vision, balance, speech, hearing, and paralysis on one side) have their own assortment of injuries in different degrees. About 25% to 40% of people with stroke acquire aphasia.

synapses. Connections that make the brain work. A synapse consists of the outgoing terminal (of the sending cell), the space (gap) between, and the incoming terminal (of the receiving cell). The outgoing terminal converts an electrical impulse to a chemical messenger (neurotransmitter) that floats across the space and transmits the message to the next cell.

therapeutic. Can be either a therapeutic drug, or a therapeutic exercise, or something considered therapeutic that helps heal or restore health. All language activities provided by speech therapists and their patients are considered "therapeutic," as well as practicing reading, writing, and speech activities while working alone or with family and friends.

video recording. Helps people with aphasia by recording their actions. They later observe what they have been doing and saying. It offers a new neurological perspective that provides the feedback to "see" what one has been unable to see (or hear) before. Feedback is a very important therapeutic facet of recovery.

walking. An important component of brain health, plasticity, and recovery, given that walking and various kinds of exercise induce a number of neurotransmitters that are particularly well suited to assist with the process of plasticity at the cellular level.

word-finding. The most common problem for people with aphasia is not being able to express a word they want to say. Instead, people with word-finding problems use empty words to fill in the missing word in a sentence, with: *that, that thing, you know, it*. There are word-finding recall strategies using *who*, *what*, *why*, *where*, and *when* questions, to help describe, write, or draw the item at the center of the hunt. Literally, the more you *search* for that word, the closer you get to the target.

About the Author

Thomas G. Broussard, Jr., Ph.D.
Three-Time Stroke Survivor
Johnny Appleseed of Aphasia Awareness

Dr. Broussard was an associate dean at The Heller School at Brandeis University until his stroke in 2011. He started Stroke Educator, Inc. (2015) and founded Aphasia Nation, Inc. (2020), a nonprofit. Both are focused on Tom's international Aphasia Awareness Campaign dedicated to educating the wider public about aphasia, a life-changing language disorder that few people have ever heard of.

As of the publication of this book (March 2022), he has given over 450 presentations—to more than 13,000 people with aphasia and their families and caregivers, as well as to clinicians and students—extending over 32 US states, plus India and Hong Kong.

To learn more, visit **StrokeEducator.com** and **AphasiaNation.org**. To contact the author regarding speaking to your group, please write to tbroussa@comcast.net, phone 207-798-1449, or contact Stroke Educator, Inc., 4 Aspen Drive, Brunswick, ME 04011.

Additional Books by Thomas G. Broussard, Jr.

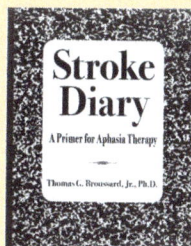

Stroke Diary: A Primer for Aphasia Therapy is practically a day-to-day diary from a stroke survivor who couldn't write … but kept on writing anyway. A first-of-its-kind primer that blazes the trail for new aphasia therapy.

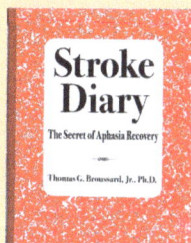

Stroke Diary: The Secret of Aphasia Recovery is a personal, intensive, enriched therapy boost for recovery, drawn from an almost 500-page diary. The secret of recovery from aphasia is all about the *doing*.

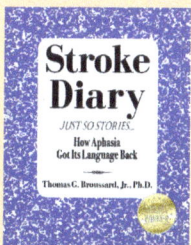

Stroke Diary: Just So Stories … How Aphasia Got Its Language Back validates how practice can provide the cure to aphasia recovery. Practice is more than just *practice*. Practice is the prescription for improvement *and* the cure.

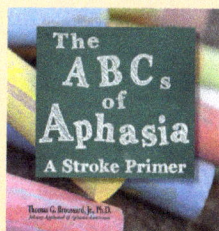

The ABCs of Aphasia: A Stroke Primer is needed by every stroke survivor and one's family immediately after a stroke. Literally, it is the A-to-Z primer about stroke, aphasia, and recovery that you can't get anywhere else.

www.ingramcontent.com/pod-product-compliance
Lightning Source LLC
Chambersburg PA
CBHW052346210326
41597CB00037B/6268